THE BLACKSMITHING INSTRUCTORS' GUIDE

THE BLACKSMITHING INSTRUCTORS' GUIDE

Sixteen lesson plans with advice on teaching techniques

DAVID HARRIES

Illustrations by Bernhard Heer

PRACTICAL ACTION
Publishing

Practical Action Publishing Ltd
The Schumacher Centre
Bourton on Dunsmore, Rugby,
Warwickshire CV23 9QZ, UK
www.practicalactionpublishing.org

© ITDG Publishing, 1993

First published in 1993
Reprinted 2001, 2005
Transferred to digital printing 2013

ISBN 978-1-85339-214-6 Paperback

A catalogue record for this book is available from the British Library.

The author has asserted his right under the Copyright Designs and Patents Act
1988 to be identified as author of this work.

Since 1974, Practical Action Publishing (formerly Intermediate Technology
Publications and ITDG Publishing) has published and disseminated books
and information in support of international development work throughout
the world. Practical Action Publishing is a trading name of Practical Action
Publishing Ltd (Company Reg. No. 1159018), the wholly owned publishing
company of Practical Action. Practical Action Publishing trades only in
support of its parent charity objectives and any profits are covenanted back
to Practical Action (Charity Reg. No. 247257, Group VAT Registration
No. 880 9924 76).

Contents

Foreword

Intermediate Technology has been working with blacksmiths in rural Africa to devise training courses which will introduce students to the use of easily accessible equipment and scrap metal so that they can make agricultural implements and other tools at an affordable cost.

The author of this book has been involved in development technology for the past eight years, and during this time has lived and worked in west and southern Africa. In the course of his work he has seen many institutes which have modelled themselves on Western training centres, using equipment and syllabuses which bear no relation to the real world in which they operate.

Here is an alternative approach, designed to help blacksmiths to identify and deliver effectively training which is appropriate to the needs of the community.

Introduction

To teach blacksmithing effectively you will need to be able to do two things. Firstly you will need to have all of the skills you intend your trainees to learn. Secondly you will need to know how to pass those skills on to others. This short guide aims to give you some useful tips on how to put your skills across effectively to your trainees. The guide is intended for all experienced blacksmiths who want to teach their skills to other people.

The method of teaching outlined in this guide is designed to provide a system of training which can be used when working with an apprentice in your own workshop, when running training courses in a rural area or when working as an instructor in a training centre. The system is structured to help you run an effective course based around carefully prepared practical demonstrations, while allowing you to bring your own style of working into each session.

The guide is divided into two sections. The first is an outline of techniques which will help you develop a systematic approach to training. Within that section are hints on how to develop your own lesson plans, how to demonstrate effectively and how to make your teaching more relevant to the needs of your trainees.

The second section is a selection of lesson plans for you to use; these are designed to enable you to develop them further in accordance with your own way of working.

PART ONE

TEACHING TECHNIQUES

STRUCTURING A TEACHING SESSION

Whenever a piece of information or a skill is to be passed on from one person to another a structured approach will make that transfer more effective. The simplest structure is to repeat the information three times: for example, a news programme on the radio will usually start off with the headlines, followed by the news in detail and finally a summary.

In this way you are first told the main points of the news, then you are given the full story and finally you are reminded of the key points of interest. The first half of this guide works in the same way; it begins with an introduction to let you know what to expect, then comes the detailed information and finally, before you get to the lesson plans, there is a brief reminder of the points covered. A good structure for teaching sessions can be produced by taking this idea and developing it.

To begin each session there must be an introduction. This can be broken down into three parts:

WHAT — Tell the trainees what the session is about. People like to know right at the start what it is they are expected to learn.

WHY — Tell the trainees what they will get out of the coming session. Will it be a new skill to help them earn money, or a new way of doing something which will save them time or materials? A trainee who feels that by listening he or she is really going to gain something will learn much more effectively than someone who is not sure why they are there.

HOW — Tell the trainees how you intend the session to be run. If you are planning to give people a test at the end of a session or to get them to make something using the skills you have just shown them, then warn them in advance. People are likely to follow things more carefully if they know what to expect from you.

If you turn back to the introduction of this guide you will see that it has been broken down into three paragraphs with a clear WHAT, WHY and HOW. The first paragraph tells you what the guide is, the second tells you why you should be interested in reading it and the final ones tell you how the guide is laid out.

After you have introduced the session, the main part of the lesson can begin. As a blacksmithing instructor your teaching sessions may be designed to show someone how to make a tool or product, or they may be designed to teach the theory of marketing. Whatever the subject, the chances are that before the session is over you will have tried to pass on to your trainees a large amount of information.

Just as it is impossible to eat and digest enough food for a whole week in one sitting, it is also impossible to digest all the information which a good teaching session might contain if it is given in one continuous block. Think back to an occasion when you have asked someone the way to a place. If you are lucky enough to have asked a good teacher, then the chances are that the instructions will have been broken down into clear stages, finishing with a reminder of the key points of the route. Afterwards they may even have questioned you to check that you have understood everything. Sadly, on most occasions people give you so much information so quickly that by the time they have finished talking to you, most of what they have said has been forgotten and it will not be long before you are looking for the next person to ask. A good teaching session is just the same; once the introduction is over, the subject of the lesson needs to be broken down into small digestible pieces of information or skills training. Each piece of information should then be checked before the next stage of the session can begin.

In the case of a practical blacksmithing session, each stage may well be in the form of a 10-minute demonstration followed by some questions to check that you have managed to get all the key points across to the trainees. The trainees should then perform the task which you have just demonstrated before the next stage can begin.

In the case of a lesson covering theory each stage may well cover just one part of a process; for example, in a session about marketing, one stage could well be about how to use hardware shops to identify possible new products. Each stage of the session should be followed by questions to check that you have put the information across clearly to the trainees.

After the main part of the session it is time to give a summary. The summary should be designed to remind the trainees of the key points of the session. Often the most effective way of summarizing is to use questions, letting the answers of the trainees form the summary. Once the summary is completed all that remains to do is to let the trainees know what the next session will cover. The following is a quick reference chart to remind you of the structure.

THE INTRODUCTION

Tell the trainees **WHAT** they are about to learn, **WHY** they should be interested and **HOW** the session will be run.

THE INSTRUCTION

Break the information down into **SMALL DIGESTIBLE STAGES**; use **QUESTIONS** to check that you have got the information across.

THE SUMMARY

Run through **KEY POINTS** and finish by telling the trainees what will be covered in the **NEXT SESSION**.

PRODUCING A LESSON PLAN

Understanding how to structure a teaching session is useful, but without careful plans it is difficult to stick to even the simplest of structures. The easiest way to make sure that your teaching sessions are properly structured is to produce a lesson plan. The plan should not only help you to run the session but should also remind you of what it is you hope to achieve and what preparation needs to be done before the trainees arrive.

A good lesson plan should contain the following:

- A name for each session and an indication of where it fits into the course.

- An estimate of the time the session is likely to take.

- A clearly stated objective.

- A list of the materials, tools and equipment needed.

- Some notes to remind you of any preparation that needs to be done before the session starts.

- A description of the What, Why and How of the session.

- An outline of the stages of the session with a list of key points for each.

- A reminder of what the next session will cover.

Lesson plans should help you to prepare for the session and to follow the structure while you are teaching. Initially, it may be advisable to make a plan for each demonstration to ensure that you have not overlooked anything. It is best to separate each stage clearly on the lesson plan and to keep your notes as brief as possible.

DEMONSTRATING TECHNIQUE

Blacksmithing is a practical activity, and therefore the most effective way to pass the skill on from one person to another is through practical demonstrations. These work best if they follow a system which tells the trainee what he or she is about to see, shows them the skill clearly and then finishes by reinforcing what has been seen. In this way the trainees leave the demonstration with the confidence to put what they have seen into practice. The following system will help you to structure your demonstrations so that you will achieve all of the aims mentioned above.

For a demonstration to work it needs to be fairly short. A good demonstration should take no more than 10 minutes from the time the trainees are gathered together to the time they start to try the skill out for themselves.

Each demonstration should start with an introduction very similar to the one used at the start of a teaching session. The trainees need to know what they are about to see, why they should be interested, how the demonstration is

going to be done and finally what safety precautions they need to remember.

After the introduction you are ready to demonstrate the skill to the trainees. Blacksmithing skills are not easy to learn: if you demonstrate a skill to the trainees just once it is unlikely that they will go away with the confidence to try that skill for themselves. One of the most effective ways to demonstrate is to show the skill three times. The first time, work silently and perform the skill at the speed you would normally expect a skilled blacksmith to work. The second time, work slowly and explain what you are doing at each stage. The final time, work again at the normal speed and emphasize the key points.

All that now remains to be done before the trainees can go and practise the skill is for you to check, using questions, that the trainees have seen and understood all the key points.

Opposite is a quick reference chart to remind you of the structure of your demonstration.

THE INTRODUCTION

Tell the trainees **WHAT** they are about to see, **WHY** they should be interested, **HOW** the demonstration will be done and what **SAFETY POINTS** need to be observed.

THE DEMONSTRATION

Perform the task once at **NORMAL** speed silently, once **SLOWLY** with a description and once at **NORMAL** speed, outlining key points.

FINAL QUESTIONS

Ask questions covering all the **KEY POINTS** before allowing the trainees to try the skill for themselves.

PLANNING A DEMONSTRATION

Until you are accustomed to using the demonstration system described on the previous page you may find it helpful to make a plan for each demonstration. Later on, all you will need is a checklist of materials, tools and equipment needed and a reminder of the key points to be included in the demonstration.

A good demonstration plan should contain the following:

- A list of the tools and equipment needed to perform the task.

- A list of the materials needed, for you and for each trainee.

- Details of what preparation needs to be done before the trainees gather round.

- An outline of the introduction including the What, Why, How of the demonstration and any safety points which need to be covered before the demonstration can take place.

- A list of key points to be covered when demonstrating, which can also be used for questions at the end.

The next three pages provide a demonstration plan layout which you can draw on when making up your own demonstration plans.

TITLE

DURATION

OBJECTIVE

By the end of the session each participant will ....
....

MATERIALS REQUIRED

FOR DEMONSTRATING

....
....
....

PER PAIR OF TRAINEES

....
....
....

TOOLS AND EQUIPMENT NEEDED

....
....
....

FINISHED PRODUCT

PREPARATION

....
....

INTRODUCTION

WHAT

I am going to demonstrate one way to ....

....

WHY

By learning this skill you will be able to ....

....

HOW

I am going to explain a few safety points. Then I will demonstrate the skill three times: once silently at the normal speed that a skilled blacksmith would do it; once slowly, explaining in detail what I am doing and finally at the normal speed. Afterwards I will ask you some questions to check that you have understood, and then you are going to have a try.

SAFETY

....

....

DEMONSTRATION 1

KEY POINTS

....

....

....

....

DEMONSTRATION 2

KEY POINTS

....

....

....

....

DEMONSTRATION 3

KEY POINTS

....

....

....

....

DEMONSTRATION 4

KEY POINTS

....

....

....

....

SUMMARY

KEY POINTS

....

....

FURTHER TRAINING

Next session we are going to ....

....

QUESTIONING TECHNIQUE

Questions can be used in a teaching session to achieve many aims. Three of the most important uses for the trainer are:

- To check whether your teaching has been effective. For example, if you have just given a demonstration you could ask the trainees questions relating to all of the key points you intended them to absorb.

- To find out how much the trainees already know about a subject before you start teaching it.

- To ensure that all of the participants are actively involved and that their ideas and thoughts are integrated into the session.

The most important thing to remember about questioning is that it should be used positively. If used badly, questions can unnerve the trainees, destroy their trust and act as a barrier between you and the group. If used well they can help to build confidence, bring out new ideas and help you develop a better understanding of what the trainees need to know and how best to put it across. The way to use questions positively is to make sure that you are trying to bring out the trainees' strengths, not their weaknesses, and to establish what they know rather than what they do not know. If you are asking questions about something you have already taught, two points worth remembering are:

- A correct answer deserves praise.

- An incorrect answer means, on most occasions, that either the question was unfair (perhaps it covered something you

had not taught) or that you have failed to teach the point effectively.

There are two systems for asking questions, each with a different purpose. The first is to put a question to the whole group. This system is useful when you are trying to get ideas from the group or when you wish to know if any of the trainees have already covered the subject you are about to teach. It can also be used to emphasize a point which you feel did not come across clearly in your teaching.

Alternatively, you may choose a member of the group to answer your question. This type of questioning is effective for checking that trainees have understood the key points after each stage of a session or after a demonstration. It can also be used to summarize a session. If used well this type of questioning will ensure that every trainee is thinking about how to answer every question.

However this will only happen if approached in the following way:

- Make sure that the group knows you are going to decide who is to answer.

- Ask the question.

- Leave plenty of time so that all the group can think of an answer.

- Choose a trainee.

Questions are among the main tools you can use as an instructor, so it is worth putting time into developing your technique.

SETTING A SYLLABUS

Unlike other trades where formal and rigid courses have been developed, blacksmithing instructors are fortunate in that most are free to develop training to suit the needs of their

trainees. This means that, when considering what to teach, you should be free to include only the skills which are of direct benefit to the participants.

Identifying these skills is not always easy. The range of skills available in text books is often very broad. It is far too easy simply to work your way through the whole range (many of which have been developed to assist in the production of decorative ironwork) rather than to identify the skills which are truly appropriate.

One system for setting up a syllabus is to look at the following four areas:

- *Materials* What types of metal and fuel will be available (and affordable) to your trainees after training? If the only available materials will be used car and truck parts, maybe the course should focus on how to get the best use out of different types of scrap metal. If coke is expensive and unavailable in some areas, maybe the course should cover the most efficient ways to produce and use charcoal.

- *Products* When your trainees have finished their training, what type of product will they be able to make successfully and market at a profit? The syllabus will have to take into account whether they will only have access to a rural market or whether they will also be able to market their products in urban areas.

- *Tools* Which blacksmithing tools are required to make the appropriate products? This is likely to be affected by factors such as whether your trainees work alone or with an assistant.

- *Processes* Which processes are required in order to make the appropriate tools and products?

A combination of talking to practising blacksmiths, looking at retail outlets and talking to families in the areas where your trainees will be working should provide answers to these questions. Only when you have answers to all these questions can you put together a syllabus that will meet the needs of your trainees.

THE LEARNING ENVIRONMENT

For many trainees, where they are taught can be as important as what they are taught. In many institutions around the world people have found that the type of workshop and the way it is equipped has a major effect on the way in which people learn. European and American training centres have been designed to reflect the conditions in which the trainees will find themselves after their training is complete. By simply copying these workshops without paying any attention to the real world which surrounds them, many training centres have created problems for themselves and their trainees. Some typical problems caused by building Western-style workshops in 'non-Western' situations are:

- On returning to their home area, trainees complain that in order to set up their own businesses they need a lot of capital to build and equip a proper workshop.

- Trainees start to think that the traditional way of working is not dignified enough for them.

- The training workshop which was expensive to set up soon becomes expensive to run as the imported tools and machines start to break down.

One of the best ways to avoid all of these problems is to take the training into the area where it is needed. This is now almost standard practice in the field of agriculture, and it has been implemented successfully in a number of rural blacksmith training projects in Africa and Asia.

However, it is often not possible to take the training out to where the blacksmiths are working. The next best thing is to try to create a learning environment which relates as closely as possible to the conditions which the trainees will find themselves in when they leave. One training centre in Zimbabwe has built a workshop using traditional methods and materials and equipped only with tools and equipment which can be made within the workshop. Not only was the workshop cheap to build, it is also a very pleasant place in which to work. The high thatched roof absorbs much of the noise, the earth floor is easy on the feet and the open sides provide good ventilation.

Your training centre should suit the conditions around you.

BUSINESS AND PRODUCTION TECHNIQUES

When teaching people blacksmithing skills it is important to introduce them to some of the production and business problems they will face when they have to use those skills to earn a living. Otherwise they may find themselves unable to put all they have learnt into practice. They may be able to produce work of a high quality but at an unrealistic cost. Even if they can make products cheaply and quickly enough, they may lack the marketing skills needed to survive.

One way in which many training centres have tried to develop these additional important skills is to incorporate training with production. This is often achieved by setting up the training centre as a production unit. In theory, this gives the trainees an understanding of production techniques and at the same time produces an income for the training centre. Though this can work well, more often than not it fails. The danger is that once a product which sells well has been found, all thoughts of training in a wide range of skills are forgotten and the centre becomes a subsidized business, often competing with the very people it was set up to help.

Another approach is to set aside special times for production days. On these days the trainees treat their work as though they were in business for themselves. The trainees work in pairs or small groups and pay for all their fuel and material.

The trainees are invited to submit tenders for various orders, which could range from 100 cold chisels to a dozen axes. It is surprising how many new ideas on production techniques are inspired by these days and how quickly trainees see the need for careful pricing and efficient use of fuel and material.

STEP-BY-STEP BOARDS

Step-by-step boards are a form of teaching aid; they help you to convey your skills and knowledge more effectively to the trainees. The boards are made up of real examples of each stage in the production of a tool or product, mounted in sequence, starting at one side with the raw materials and ending up at the other side with an example of the finished product. Thus trainees can see how each stage in production follows on from another. For example, a step-by-step board showing the stages in the production of a hammer head would be laid out as follows from left to right:

Stage 1 Half-shaft including splined section.
Stage 2 Half-shaft with splined section cut off.
Stage 3 Half-shaft with peen drawn down on the end.
Stage 4 Half-shaft with peen and slot cut through.
Stage 5 Half-shaft with peen and eye drifted through.
Stage 6 Hammer head ready to cut from shaft.
Stage 7 Hammer head ready for hardening and tempering.
Stage 8 Hammer head showing final temper colours.

Each item in the display has to be mounted on the boards. This can be done by welding a short bolt, out of sight, on the back of the workpiece, so that each stage can be held firmly in place. All the metal parts should be varnished to protect them against rust.

Constructing a board can clearly involve a lot of work and material; however, once made, a good board will last many years. A well-made board can be so effective that trainees are able to make a product just by copying what they see.

SUMMARY

How can using a structure improve the way trainees learn?

A structured approach helps you to put your skills and knowledge across in a systematic way when teaching. It ensures that you give information in digestible portions and helps you to check that the trainees are indeed learning.

What are the main benefits of producing and using a session plan?

It will help you to prepare for a teaching session and to keep to a structure; it will also remind you of the key points you need to cover.

What are the key points when demonstrating?

Keep the demonstration short; outline any safety points; where possible run through the skill three times and remember to check that the key points have been seen and understood by the trainees before they practise the skill for themselves.

Does every demonstration need a plan?

Once you are used to the system all you will need is a list of what to prepare and an outline of key points.

What are the main uses of questions?

To check your teaching, to assess the trainees' knowledge and to encourage participation. Remember to use questions positively.

What should I take into account when setting a syllabus?

Any syllabus should be based on the skills that your trainees will need in order to succeed after they have been trained. It should take into account the tools, equipment and raw material they will have access to in the 'real world' in which they will be working.

Does it matter where the training takes place?

Yes, the training environment should reflect the environment in which the trainees are going to work.

What benefits are there in bringing production ideas into the training?

All skills training needs to be backed up with an understanding of how to adapt those skills in order to make a living. Bringing production and marketing ideas into your sessions will help you to do this.

How will the use of step-by-step boards help the trainees?

They can help trainees who have missed a session to catch up without the need for much support. They will help others to experiment in making things in their own time and they will encourage thought and discussion about systems of production.

PART TWO

LESSON PLANS

DRAWING DOWN

DURATION 30 MINUTES

OBJECTIVE

By the end of the session each participant will have drawn down a piece of mild steel to a round point.

MATERIALS REQUIRED

FOR DEMONSTRATING

Three 200mm lengths of 12-15mm round section mild steel bar.
Sample of finished product.

TOOLS AND EQUIPMENT NEEDED

Anvil
Hearth
Bellows
Hand hammer
Hollow-bit tongs

PER PAIR OF TRAINEES

Two 200mm lengths of 12-15mm round section mild steel bar.

FINISHED PRODUCT

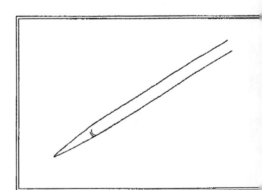

PREPARATION

Heat all three pieces of steel up to a dull red before gathering the trainees together for the first demonstration.

| WHAT | Draw down a round point on a piece of mild steel bar (pass round sample of finished item). |

| WHY | Drawing down is a key skill which you will use when making many tools and products e.g. tangs on axes and hoes. |

| HOW | I shall demonstrate each stage, asking some questions, then each of you will draw down a round point. |

DEMONSTRATION 1

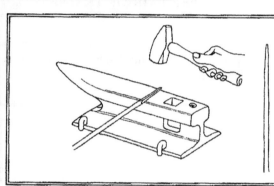

KEY POINTS

Check that your tongs fit.
Yellow heat.
Start at the tip.
Forge down to a square point.
Work near edge of anvil.
Keep work at correct angle.
Avoid cold working the steel.

DEMONSTRATION 2

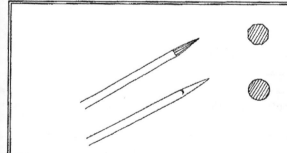

KEY POINTS

Yellow heat (warn about danger of the tip burning).
Work at edge of anvil.
Remove corners, working along ridges.

FURTHER TRAINING

The next exercise will be the use of a hot set.

HOT CUTTING

DURATION 30 MINUTES

OBJECTIVE

By the end of the session each participant will have used a hot set to cut through a piece of mild steel.

MATERIALS REQUIRED

FOR DEMONSTRATING	PER PAIR OF TRAINEES
Three pieces of mild steel suitable for cutting.	Two pieces of mild steel suitable for cutting.

TOOLS AND EQUIPMENT NEEDED

Anvil
Hearth
Bellows
Sledgehammer
Hot set

PREPARATION

Heat all three pieces of steel up to a dull red before gathering the trainees together for the first demonstration.

WHAT

Use a hot set to cut through a piece of mild steel.

WHY

Hot cutting is a key skill which you will use when making many tools and products e.g. cutting out axe blanks.

HOW

I shall demonstrate each stage, asking some questions, then each of you will use a hot set to cut up some pieces of scrap steel.

DEMONSTRATION 1

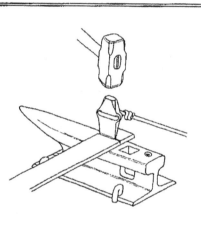

KEY POINTS

Do not confuse hot set with cold set.
Quench set every few blows.
Correct use of sledgehammer.
Do not cut right through metal.
Work at correct forging temperature.

FURTHER TRAINING

The next exercise will be to upset a piece of mild steel bar.

UPSETTING

DURATION 30 MINUTES

OBJECTIVE

By the end of the session each participant will have upset a piece of mild steel.

MATERIALS REQUIRED

FOR DEMONSTRATING

Three 400mm lengths of 12-15mm round section mild steel bar.
Sample of finished product.

PER PAIR OF TRAINEES

Two 400mm lengths of 12-15mm round section mild steel bar.

TOOLS AND EQUIPMENT NEEDED

Anvil
Hearth
Bellows
Hand hammer

PREPARATION

Heat all three pieces of steel up to a dull red before gathering the trainees together for the first demonstration.

WHAT

Upset the end of a piece of mild steel bar (pass round sample of finished item).

WHY

Upsetting is a skill which you can use when you wish to increase the thickness of a piece of metal, e.g. when preparing for a forge weld.

HOW

I shall demonstrate, asking some questions, then each of you will upset the end of a piece of mild steel.

DEMONSTRATION 1

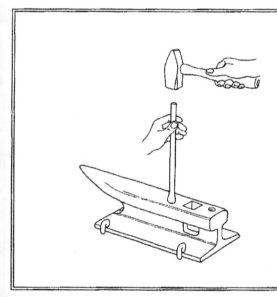

KEY POINTS

Work at near welding heat.
Use of water to control heat.
Correct any bending immediately.

FURTHER TRAINING

Next session we are going to make a round punch.

MAKING A ROUND PUNCH

DURATION 2 HOURS

OBJECTIVE

By the end of the session each participant will have made a round punch.

MATERIALS REQUIRED

FOR DEMONSTRATING

Three vehicle coil springs.
Piece of chalk.
Sample of finished product.

PER PAIR OF TRAINEES

One vehicle coil spring.
Piece of chalk.

TOOLS AND EQUIPMENT NEEDED

Anvil
Hearth
Bellows
Hand hammer
Hollow-bit tongs
Sledgehammer
Hot set
Quenching trough

FINISHED PRODUCT

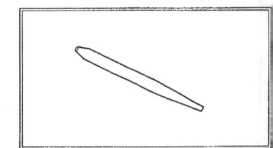

PREPARATION

Have two of the coil springs already set correctly in the fire and heating up before the trainees are gathered for the first demonstration.

WHAT	Make a round punch from a piece of vehicle coil spring (pass round sample of finished product).

WHY	This is a tool which you can keep and will find very useful when punching holes, for example when repairing hoes.

HOW	I shall demonstrate each stage, asking some questions, then each of you will make a round punch.

DEMONSTRATION 1

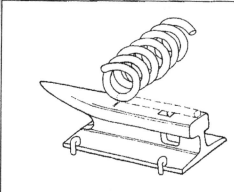

Marking off the metal

KEY POINTS

Chalk a line 200mm from end of anvil.
Mark out length on spring.
Place carefully in fire, chalk line in hottest part.

DEMONSTRATION 2

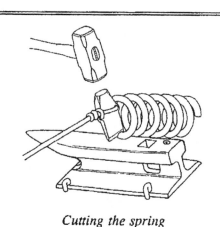

Cutting the spring

KEY POINTS

Explain difference between hot and cold sets.
Importance of quenching.
Work at correct forging temperature.
Correct use of sledgehammer.
Keeping clear of sledgehammer.
Danger of spoiling anvil.
Use of tongs to break off metal at cut.

DEMONSTRATION 3

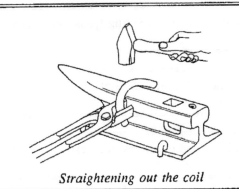

Straightening out the coil

KEY POINTS

Correct positioning in fire.
Use of hardie hole to start the straightening process.
Turn workpiece around in tongs.
Check for straightness.

DEMONSTRATION 4

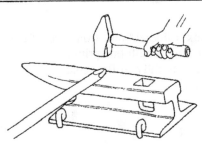

Forging the striking end

KEY POINTS

Draw down a short taper.
Work near edge of anvil.
Flatten back end.

DEMONSTRATION 5

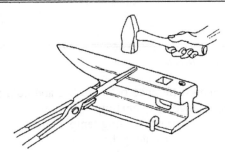

Drawing down the punch

KEY POINTS

Draw down square to a gentle taper.
Work near edge of anvil.
Remove corner ridges.
Round off carefully.
Check for straightness.

FURTHER TRAINING

Next session we are going to make a hot chisel from a piece of half-shaft.

MAKING A HOT CHISEL

DURATION 3 HOURS

OBJECTIVE

By the end of the session each participant will have made a hot chisel from a length of half-shaft or anti-roll bar.

MATERIALS REQUIRED

FOR DEMONSTRATING

Used half-shaft or anti-roll bar from a car or light van.
Sample of finished product.
Piece of chalk.

PER PAIR OF TRAINEES

Used half-shaft or anti-roll bar from a car or light van.
Piece of chalk.

TOOLS AND EQUIPMENT NEEDED

Anvil
Hearth
Bellows
Hand hammer
Hollow-bit tongs
Sledgehammer
Hot set
Quenching trough
Old file (for hot-filing)

FINISHED PRODUCT

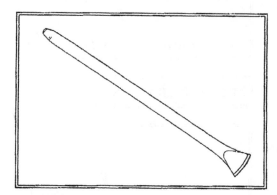

PREPARATION

Heat up the splined section of the shaft before gathering the trainees together for the first demonstration.

WHAT

Make a hot chisel from a length of half-shaft or anti-roll bar (pass round sample of finished product).

WHY

This will be another tool for you to add to your tool-kit; if you have to work alone you will find this tool easier to use than the hot set.

HOW

I shall demonstrate each stage, asking some questions, then each of you will make your own hot chisel.

DEMONSTRATION 1

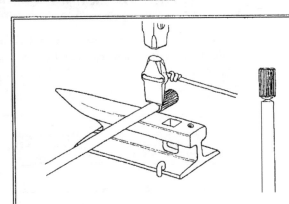

Cutting off the splined section

KEY POINTS

Work at correct forging temperature.
Turn work and cut towards centre of bar.
Quench set.
Finish cut over edge of anvil.

DEMONSTRATION 2

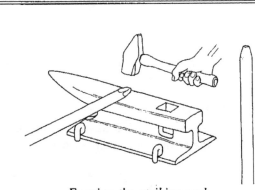

Forging the striking end

KEY POINTS

Draw down short taper.
Work near edge of anvil.
Flatten back end.

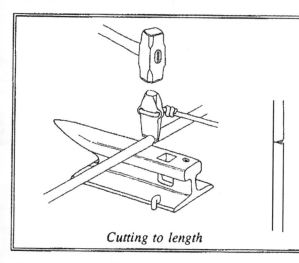

KEY POINTS

Mark off length (400mm approx.).
Place mark at hottest part of fire.
Work at correct forging temperature.
Cut towards centre by turning bar.
Finish cut over edge of anvil.

Cutting to length

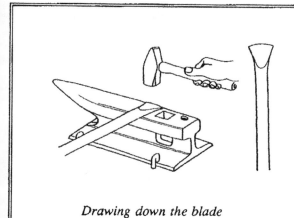

KEY POINTS

Work at correct forging temperature.
Work near edge of anvil.
Allow blade to spread slightly.
Hot-file the cutting edge.

Drawing down the blade

FURTHER TRAINING

Next session we are going to make a cold chisel from a piece of coil spring.

MAKING A COLD CHISEL

DURATION 4 HOURS

OBJECTIVE

By the end of the session each participant will have made a cold chisel.

MATERIALS REQUIRED

FOR DEMONSTRATING

Three vehicle coil springs.
Piece of chalk.
Sample of finished product.

PER PAIR OF TRAINEES

One vehicle coil spring.
Piece of chalk.

TOOLS AND EQUIPMENT NEEDED

Anvil
Hearth
Bellows
Hand hammer
Hollow-bit tongs
Sledgehammer
Hot set
Quenching trough
Buckets or cans (for shallow quenching)
File
Piece of old grinding wheel

FINISHED PRODUCT

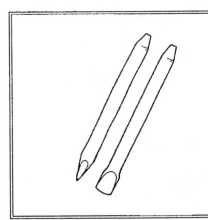

PREPARATION

Have two of the coil springs already set correctly in the fire and heating up before the trainees are gathered for the first demonstration.

WHAT	Make a cold chisel from a piece of vehicle coil spring (pass round sample of finished product).

WHY	This is a tool you can keep and which you will find very useful when cutting cold metal, for example when cutting out hoe blades.

HOW	I shall demonstrate each stage, asking some questions, then each of you will make a cold chisel.

DEMONSTRATION 1

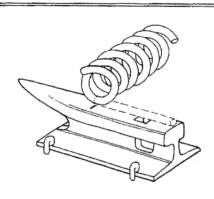

Marking off the metal

KEY POINTS

Chalk line 200mm from end of anvil.
Mark out length on spring.
Place carefully in fire, chalk line in hottest part.

DEMONSTRATION 2

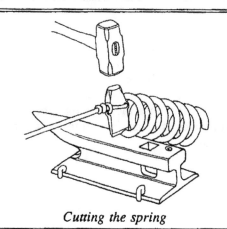

Cutting the spring

KEY POINTS

Explain difference between hot and cold sets.
Importance of quenching.
Work at correct forging temperature.
Correct use of sledgehammer.
Keeping clear of sledgehammer.
Danger of spoiling anvil.
Use of tongs to break off metal at cut.

DEMONSTRATION 3

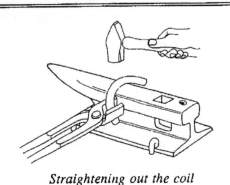

Straightening out the coil

KEY POINTS

Correct positioning in fire.
Use of hardie hole to start the straightening process.
Turn workpiece around in tongs.
Check for straightness.

DEMONSTRATION 4

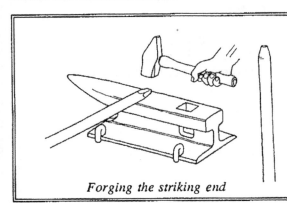

Forging the striking end

KEY POINTS

Draw down a short taper.
Work near edge of anvil.
Flatten back end.

DEMONSTRATION 5

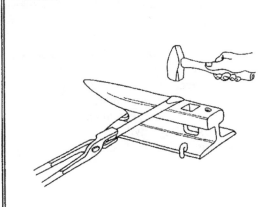

Drawing down the blade

KEY POINTS

Work at correct forging temperature.
Work near edge of anvil.
Do not allow blade to spread out.
Check chisel for straightness.
Finish off with a file, making sure that metal workpiece is clean enough to show temper colours.

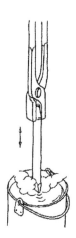

Hardening and tempering

KEY POINTS

Importance of not overheating chisel.
Quench tip (moving to avoid a shock line).
Clean metal to show temper colours.
Keep chisel away from cold anvil.
Move quickly once correct temper colour
(dark straw) is at tip.
Test finished product.

FURTHER TRAINING

Next session we are going to make a hot set from a piece of leaf spring and some round section mild steel bar.

MAKING A HOT SET

DURATION 6 HOURS

OBJECTIVE

By the end of the session each participant will have made a hot set from a piece of leaf spring and a length of mild steel rod.

MATERIALS REQUIRED

FOR DEMONSTRATING

A length of 10-12mm round section mild steel rod (approx. 1000mm).
A short length of fairly thick leaf spring.

PER PAIR OF TRAINEES

Two lengths of 10-12mm round section mild steel rod (approx. 1000mm).
Two lengths of fairly thick leaf spring.

TOOLS AND EQUIPMENT NEEDED

Anvil
Hearth
Bellows
Hand hammer
Sledgehammer
Hot set
Round punch
Old file (for hot-filing)

FINISHED PRODUCT

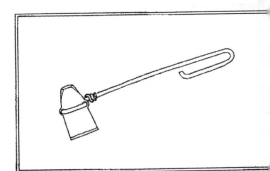

PREPARATION

Have the length of leaf spring set in the fire so that it is heating up approx. 40mm from the end before the trainees are gathered for the first demonstration.

WHAT	Make a hot set from a short length of leaf spring and some round section mild steel bar.
WHY	As you know from earlier sessions the hot set is a very useful tool. In this session you will be making one which you can keep along with the other tools you have made.
HOW	I shall demonstrate each stage, asking some questions, then each of you will make your own hot set.

DEMONSTRATION 1

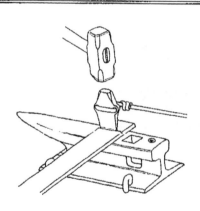

Cutting through the spring

KEY POINTS

Work at correct forging temperature.
Cut a little over halfway through spring, approx. 40mm from the end.
Quench set.
Allow metal to cool slightly and break off over edge of anvil.

DEMONSTRATION 2

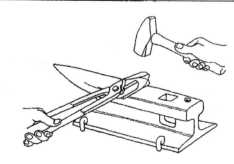

Shaping the tool head

KEY POINTS

Forge sides flat.
Work at correct forging temperature.
Forge into a wedge shape.
Flatten top and remove corners.

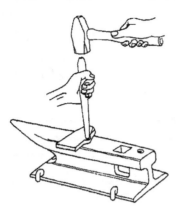

KEY POINTS

Work at correct forging temperature.
Punch three-quarters of the way through on
one side.
Turn over and drive punch through until hole
is large enough to fit handle.
Punch second hole in same way.

Punching holes for the handle

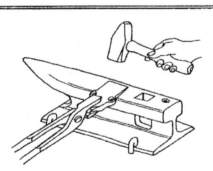

KEY POINTS

Work near edge of anvil.
Work on both sides.
Check that blade edge is on centre line of
tool.

Drawing down the blade

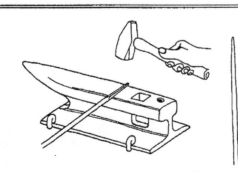

KEY POINTS

Draw down square first, then round off.
Warn about how easily the thin metal can
burn.
Bend into U-shape so that handle can be fitted
to blade section.

Drawing down the handle

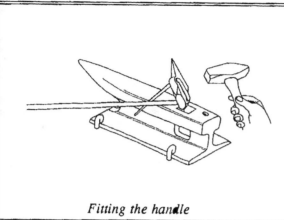

KEY POINTS

Work at correct forging temperature.
Work quickly.
Try to do the whole operation in two heats.

Fitting the handle

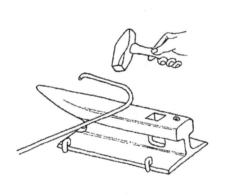

KEY POINTS

Work at correct forging temperature.
First bend approx. 15mm from end.
Second bend far enough from end to form a
comfortable handle.
Finally hot-file or grind blade to shape.

Forming the end of the handle

FURTHER TRAINING

Next session we are going to make a cold set.

MAKING A COLD SET

DURATION 6 HOURS

OBJECTIVE

By the end of the session each participant will have made a cold set from a piece of leaf spring and a length of mild steel rod.

MATERIALS REQUIRED

FOR DEMONSTRATING

A length of 10-12mm round section mild steel rod (approx. 1000mm).
A short length of thick leaf spring.
Sample of finished product.

PER PAIR OF TRAINEES

Two lengths of 10-12mm round section mild steel rod (approx. 1000mm).
Two short lengths of fairly thick leaf spring.

TOOLS AND EQUIPMENT NEEDED

Anvil
Hearth
Bellows
Hand hammer
Sledgehammer
Hot set
Round punch
Old file (for hot-filing)

FINISHED PRODUCT

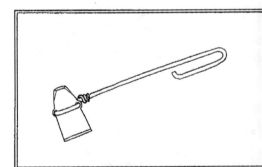

PREPARATION

Have the length of leaf spring set in the fire so that it is heating up approx. 40mm from the end before the trainees are gathered for the first demonstration.

| WHAT | Make a cold set from a short length of leaf spring and some round section mild steel bar. Hand round the sample of the finished product. |

| WHY | The cold set is very useful for cutting mild steel bar without heating up the metal. In this session you will be making one which you can keep along with the other tools you have made. |

| HOW | I shall demonstrate each stage, asking some questions, then each of you will make your own cold set. |

DEMONSTRATION 1

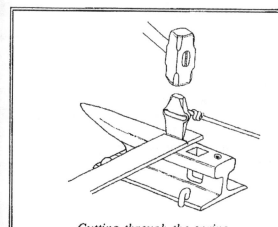

Cutting through the spring

KEY POINTS

Work at correct forging temperature.
Cut a little over halfway through the spring, approx. 40mm from the end.
Quench set.
Allow metal to cool slightly and break off over edge of anvil.

DEMONSTRATION 2

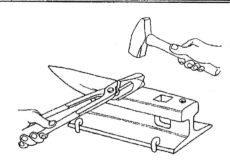

Shaping the tool head

KEY POINTS

Forge sides flat.
Work at correct forging temperature.
Forge into a wedge shape.
Flatten top and remove corners.

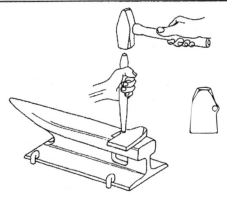

KEY POINTS

Work at correct forging temperature.
Punch three-quarters of the way through on one side.
Turn over and drive punch through until hole is large enough to fit handle.
Punch second hole in same way.

Punching holes for the handle

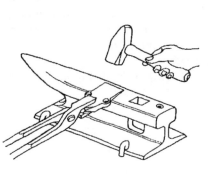

KEY POINTS

Work near edge of anvil.
Work on both sides.
Leave blade a little thicker than on hot set you made in the last session.
Check that blade edge is on centre line of tool.

Drawing down the blade

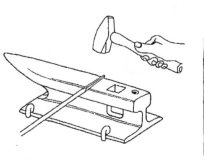

KEY POINTS

Draw down square first, then round off.
Warn about how easily the thin metal will burn.
Bend into U-shape so that handle can be fitted to blade section.

Drawing down the handle

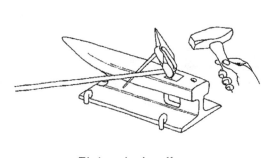

KEY POINTS

Work at correct forging temperature.
Work quickly.
Try to do the whole operation in two heats.

Fitting the handle

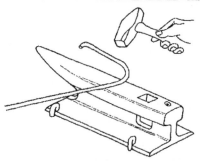

KEY POINTS

Work at correct forging temperature.
First bend approx. 15mm from end.
Second bend far enough from end to form a
comfortable handle.
File or grind blade to shape.

Forming the end of the handle

KEY POINTS — Hardening and tempering

Importance of not overheating the set.
Quench tip (moving to avoid a shock line).
Clean metal to show temper colours.
Keep set away from cold anvil.

Move quickly once correct temper (dark
straw) colour is at tip.
Test finished product.

FURTHER TRAINING

Next session we are going to make a pair of tongs.

MAKING A PAIR OF TONGS

DURATION 6 HOURS

OBJECTIVE

By the end of the session each participant will have made a pair of light tongs.

MATERIALS REQUIRED

FOR DEMONSTRATING

One 100mm length of 16mm round mild steel bar.
Short length of 10 or 12mm round mild steel bar.
Sample of finished product.

PER PAIR OF TRAINEES

Two 100mm lengths of 16mm round mild steel bar.
Two short lengths of 10 or 12mm round mild steel bar.

TOOLS AND EQUIPMENT NEEDED

FINISHED PRODUCT

Hand hammer
Hot set
Hollow-bit tongs
Round punch
File
Anvil
Hearth
Bellows
Quenching trough

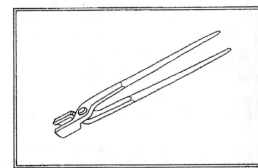

PREPARATION

As a pair of tongs is made up of two identical halves it is useful to demonstrate each operation twice, the first time at normal speed and the second slowly outlining key points.

38 *Tongs*

WHAT	Make a pair of tongs for light work out of a mild steel bar.

WHY	This pair of tongs will be a tool that you can add to your tool-kit. The pair you are making today will be suitable for holding leaf spring when making axes and adzes.

HOW	I shall demonstrate each stage, asking some questions, then each of you will make a pair of tongs.

DEMONSTRATION 1

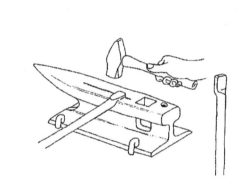

KEY POINTS

Work on nearest edge of anvil.
Work at correct forging temperature.
Work only on rounded edge of anvil.

Forming the jaws

DEMONSTRATION 2

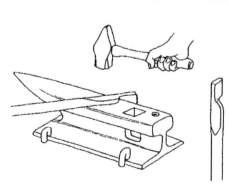

KEY POINTS

Turn a quarter of a revolution anti-clockwise (clockwise if left-handed).
Work on far side of the anvil.
Work only on rounded edge of anvil.

Forming the pivot point

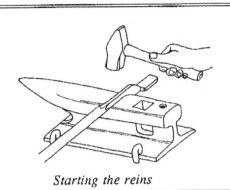

KEY POINTS

Turn work again through a quarter of a turn anti-clockwise (clockwise if left-handed).
Work over far edge of anvil.
Work only on rounded edge.
Turn metal again through a quarter of a turn and forge down to a square section.

Starting the reins

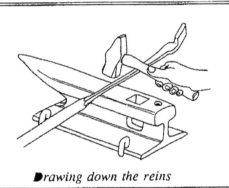

KEY POINTS

Work along the length evenly.
Draw down to a square first then round off handle part.
Cut in half at centre.
Finish off both reins to equal length.

Drawing down the reins

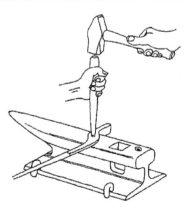

KEY POINTS

Work at correct forging temperature.
Start hole in centre of pivot area.
Punch through over a hole.
Holes in both halves must be same size.

Punching the pivot hole

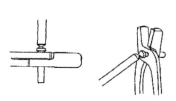

KEY POINTS

Forming the shoulder.
Check length and thickness.
Remember not to cut rivet completely off bar.
Work at correct forging temperature.
Work quickly when fitting rivet.

Making and fitting the rivet

DEMONSTRATION 7

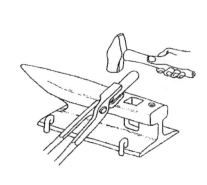

KEY POINTS

Work at correct forging temperature.
Align jaws and set to fit common size of leaf spring.
Check reins are in line and quench out, opening and closing tongs as they cool.

Finishing the tongs

FURTHER TRAINING

Next session we are going to make a hammer eye chisel which is one of the tools you will need in order to make your own hammers.

MAKING A HAMMER EYE CHISEL

DURATION 3 HOURS

OBJECTIVE

By the end of the session each participant will have made a hammer eye chisel from a length of half-shaft.

MATERIALS REQUIRED

FOR DEMONSTRATING

Used half-shaft.
Sample of finished product.
Piece of chalk.

PER PAIR OF TRAINEES

Used half-shaft.
Piece of chalk.

TOOLS AND EQUIPMENT NEEDED

Anvil
Hearth
Bellows
Hand hammer
Hollow-bit tongs
Sledgehammer
Hot set
Quenching trough
Old file (for hot-filing)

FINISHED PRODUCT

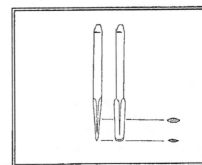

PREPARATION

Heat up the splined section of the shaft before gathering the trainees together for the first demonstration.

WHAT	Make a hammer eye chisel from a length of half-shaft (pass round sample of finished product).

WHY	This will be another tool for you to add to your tool-kit; it will be useful to you when making any type of hand hammer.

HOW	I shall demonstrate each stage, asking some questions, then each of you will make your own hammer eye chisel.

DEMONSTRATION 1

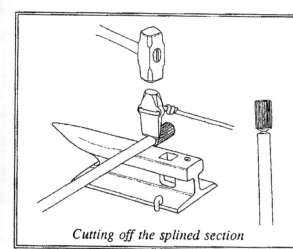

KEY POINTS

Work at correct forging temperature.
Turn work and cut towards centre of bar.
Quench set.
Finish cut over edge of anvil.

Cutting off the splined section

DEMONSTRATION 2

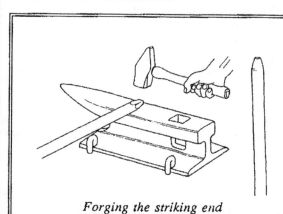

KEY POINTS

Draw down short taper.
Work near edge of anvil.
Flatten back end.

Forging the striking end

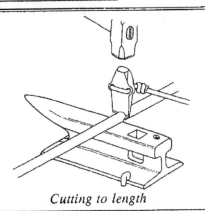

Cutting to length

KEY POINTS

Mark off length (400mm approx.).
Place mark at hottest part of fire.
Work at correct forging temperature.
Cut towards centre by turning bar.
Finish cut over edge of anvil.

DEMONSTRATION 4

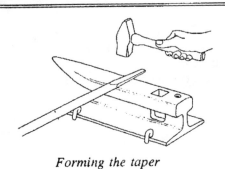

Forming the taper

KEY POINTS

Draw down blade section to a rectangular taper.
Work at correct forging temperature.

DEMONSTRATION 5

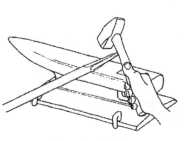

Forming the blade

KEY POINTS

Work at correct forging temperature.
Work near edge of anvil.
Forge blade leaving centre of chisel fairly thick and sides rounded.
Sharpen blade with a file.

FURTHER TRAINING

Next session we are going to make a hammer eye drift from a piece of used half-shaft.

MAKING A HAMMER EYE DRIFT

DURATION 3 HOURS

OBJECTIVE

By the end of the session each participant will have made a hammer eye drift from a length of half-shaft.

MATERIALS REQUIRED

FOR DEMONSTRATING

Used half-shaft.
Sample of finished product.
Piece of chalk.

PER PAIR OF TRAINEES

Used half-shaft.
Piece of chalk.

TOOLS AND EQUIPMENT NEEDED

Anvil
Hearth
Bellows
Hand hammer
Hollow-bit tongs
Sledgehammer
Hot set
Quenching trough
Old file (for hot-filing)

FINISHED PRODUCT

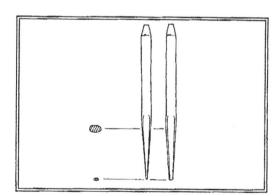

PREPARATION

Heat up the splined section of the shaft before gathering the trainees together for the first demonstration.

WHAT	Make a hammer eye drift from a length of half-shaft (pass round sample of finished product).

WHY	This will be another tool for you to add to your tool-kit; it will be useful to you when making any type of hand hammer.

HOW	I shall demonstrate each stage, asking some questions, then each of you will make your own hammer eye drift.

DEMONSTRATION 1

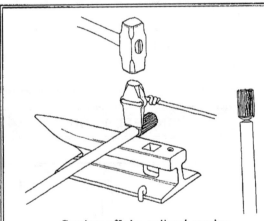

Cutting off the splined section

KEY POINTS

Work at correct forging temperature.
Turn the work and cut towards centre of bar.
Quench set.
Finish cut over edge of anvil.

DEMONSTRATION 2

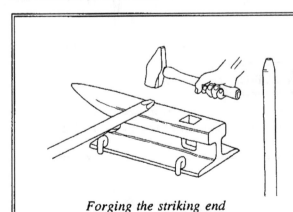

Forging the striking end

KEY POINTS

Draw down short taper.
Work near edge of anvil.
Flatten back end.

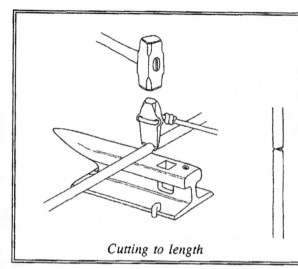

Cutting to length

KEY POINTS

Mark off length (400mm approx.).
Place mark at hottest part of fire.
Work at correct forging temperature.
Cut towards centre by turning bar.
Finish cut over edge of anvil.

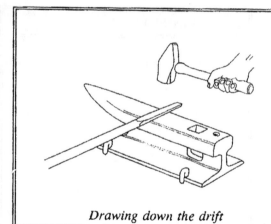

Drawing down the drift

KEY POINTS

Work at correct forging temperature.
Work near edge of anvil.
Draw down drift to a gently tapered rectangle.
Finally round off sides to form an oval.

FURTHER TRAINING

Next session we are going to make a cross-peen hammer from a piece of used half-shaft.

MAKING A CROSS-PEEN HAMMER

DURATION 8 HOURS

OBJECTIVE

By the end of the session each participant will have made a cross-peen hammer.

MATERIALS REQUIRED

FOR DEMONSTRATING	PER PAIR OF TRAINEES
Used half-shaft. Sample of finished product.	Used half-shaft.

TOOLS AND EQUIPMENT NEEDED

Anvil
Hearth
Bellows
Hand hammer
Flat-bit tongs
Sledgehammer
Hot set
Quenching trough
Old file (for hot-filing)
Can with small hole in base (for hardening)
Wooden block or tree stump

FINISHED PRODUCT

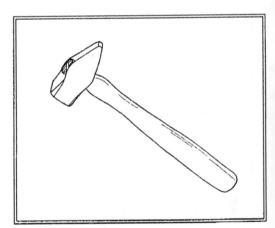

PREPARATION

Heat up the splined section of the shaft before gathering the trainees together for the first demonstration.

INTRODUCTION

WHAT

Make a cross-peen hammer from a length of half-shaft (pass round sample of finished product).

WHY

Hammer-making is a very useful skill. Once you have learnt how to make a cross-peen hammer it will be easy for you to make a wide range of hammers, some for you to use and some to sell to carpenters, builders etc. And of course it will be another tool for you to add to your tool-kit.

HOW

I shall demonstrate each stage, asking some questions, then each of you will make your own hammer.

DEMONSTRATION 1

KEY POINTS

Work at correct forging temperature.
Turn the work and cut towards centre of bar.
Quench set.
Finish cut over edge of anvil.

Cutting off the splined section

DEMONSTRATION 2

KEY POINTS

Draw down short wedge-shaped taper on end of shaft.
Work near edge of anvil.

Drawing down the peen

DEMONSTRATION 3

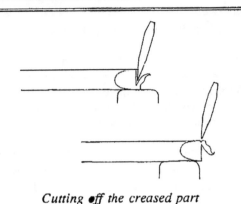

Cutting off the creased part

KEY POINTS

Work at correct forging temperature.
Cut towards centre from one side.
Turn bar over and cut into centre from other side.
Finish cut over edge of anvil.
Remember to adjust angle of set to get a straight cut.

DEMONSTRATION 4

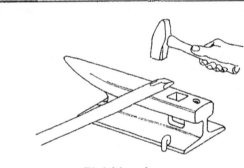

Finishing the peen

KEY POINTS

Work at correct forging temperature.
Work near edge of anvil.
Forge in corners and round off peen.

DEMONSTRATION 5

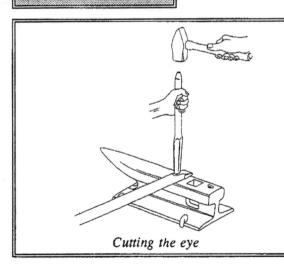

Cutting the eye

KEY POINTS

Form the flat first.
Be sure to start cut exactly in line with centre of shaft.
Work at correct forging temperature.
Quench cutting edge of chisel often.
Cut two-thirds of the way through from one side, turn over and finish cut from the other side.

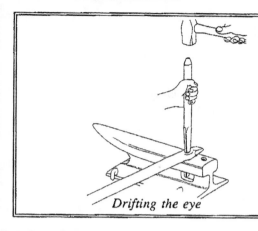

KEY POINTS

Work at correct forging temperature.
Do not quench the drift.
Drift the eye until slightly smaller than
required finished size.

Drifting the eye

KEY POINTS

Drive drift firmly into place.
Work quickly first on one side then the other.
Remove drift and drive in from the other side;
finish forging cheeks.

Forging out the cheeks

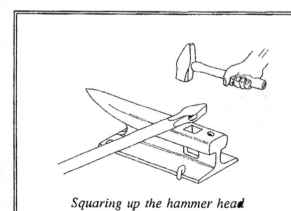

KEY POINTS

Use firm level blows.
Work at correct forging temperature.

Squaring up the hammer head

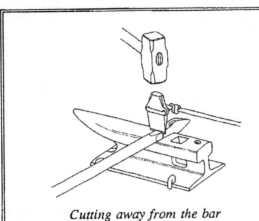

KEY POINTS

Cut evenly from all sides into centre.
Work at correct forging temperature.
Quench set frequently.
Make the last cut over edge of anvil.

Cutting away from the bar

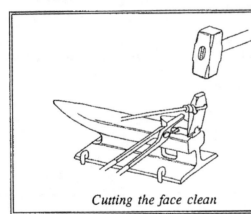

KEY POINTS

Angle set to get a straight cut.
Cut over edge of anvil.

Cutting the face clean

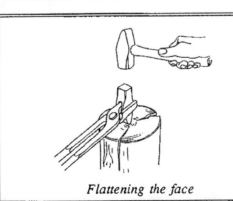

KEY POINTS

Work at correct forging temperature.
Work on wooden block to avoid damaging the peen.
Anneal finished head and file to shape when cold.

Flattening the face

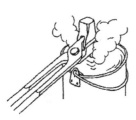

Hardening and tempering

KEY POINTS

Have plenty of cold water in the trough and a stone for cleaning the metal so that temper colours can be seen.
Do not heat above red heat.
Work quickly.
Quench face and peen as shown.
Re-quench face and peen as soon as correct temper colours are seen.

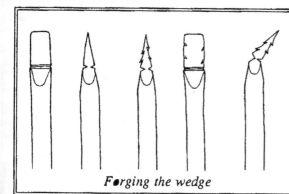

Forging the wedge

KEY POINTS

Draw down a flat-sided taper.
Work at correct forging temperature.
Cut part of the way through from both sides.
Cut barbs into wedge.
Break off wedge with a pair of tongs.
Be careful of sharp edges on barbs.

Fitting the hammer head

KEY POINTS

Make a wooden wedge.
Cut slot in shaft.
Shape shaft to fit the eye.
Drive shaft firmly into head.
Fit wooden wedge.
Fit metal wedge.

FURTHER TRAINING

Next session we are going to make an axe blade from a piece of leaf spring.

MAKING A TRADITIONAL PATTERN AXE

DURATION 3 HOURS

OBJECTIVE

By the end of the session each participant will have made a traditional pattern axe from a piece of leaf spring.

MATERIALS REQUIRED

FOR DEMONSTRATING

Two pieces of leaf spring from a car.
Sample of finished product.

PER PAIR OF TRAINEES

Piece of leaf spring from a car.

TOOLS AND EQUIPMENT NEEDED

Hand hammer
Sledgehammer
Flat-bit tongs
Hot set
File
Anvil
Hearth
Bellows

FINISHED PRODUCT

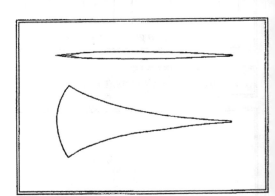

PREPARATION

Have two pieces of leaf spring already set correctly in the fire and heating up before the trainees are gathered for the first demonstration.

WHAT

Make a traditional pattern axe from a piece of used vehicle leaf spring.

WHY

This is one of the products which you can make and sell. Having made this axe it will be easy for you to make different styles of axe to suit customers in your area.

HOW

I shall demonstrate each stage, asking some questions, then each of you will make an axe like this one (hand round sample of finished product).

DEMONSTRATION 1

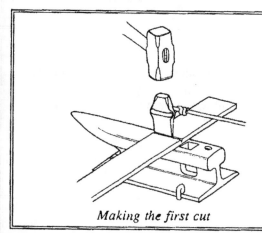

KEY POINTS

Work at correct forging temperature.
Quench set every few blows.
Cut two thirds of the way through, allow to cool and break off.

Making the first cut

DEMONSTRATION 2

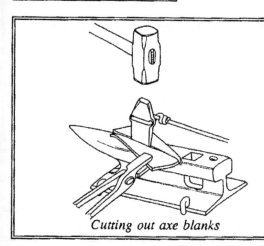

KEY POINTS

Work at correct forging temperature.
Quench set every few strikes.
Allow to cool slightly before breaking in half.

Cutting out axe blanks

Removing the corners

KEY POINTS

Work at correct forging temperature.
Remove larger piece from the longest side.
Quench corners.
Break off over edge of anvil.

Forging the blade

KEY POINTS

Work evenly along whole length of blade.
Work at correct forging temperature.
Be careful not to burn the thin metal at blade's edge.

Drawing down the tang

KEY POINTS

Forge in sides of axe until correct width is reached.
Draw down tip to form tang.

FURTHER TRAINING

Next session we are going to make a knife from a piece of leaf spring.

MAKING A SIMPLE KNIFE

DURATION 4 HOURS

OBJECTIVE

By the end of the session each participant will have made a knife from a piece of leaf spring.

MATERIALS REQUIRED

FOR DEMONSTRATING	PER PAIR OF TRAINEES
Two pieces of fairly thin leaf spring from a car. Sample of finished product.	Piece of fairly thin leaf spring from a car.

TOOLS AND EQUIPMENT NEEDED

FINISHED PRODUCT

Hand hammer
Flat-bit tongs
Sledgehammer
Hot set
File
Piece of old grinding wheel
Anvil
Hearth
Bellows
Quenching trough

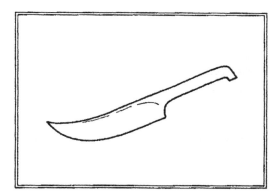

PREPARATION

Have two pieces of leaf spring already set correctly in the fire and heating up before the trainees are gathered for the first demonstration.

WHAT	Make a large knife from a piece of used vehicle leaf spring.

WHY	This is one of the products which you can make and sell. Having made this knife it will be easy for you to make different styles of knife to suit customers in your area.

HOW	I shall demonstrate each stage, asking some questions, then each of you will make a knife like this one (hand round sample of finished product).

DEMONSTRATION 1

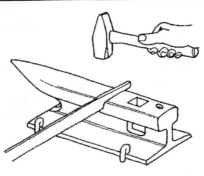

Forging the handle

KEY POINTS

Work at correct forging temperature.
Make handle slightly longer than is wanted so that the end can be bent over.
Handle should be slightly thicker than blade section and rounded off so that it will be comfortable to hold.

DEMONSTRATION 2

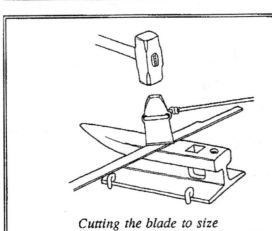

Cutting the blade to size

KEY POINTS

Work at correct forging temperature.
Quench set every few strikes.

KEY POINTS

Be careful not to allow thinner metal at the
end to burn when heating.
Work at correct forging temperature.
Curve blade over its whole length towards the
side which will become its cutting edge
(forging the blade will take out this curve).

Preparing to forge the blade

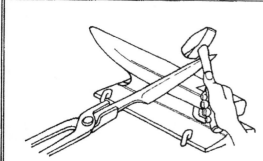

KEY POINTS

Work evenly along the whole length of the
blade.
Work at correct forging temperature.
Be careful not to burn the thin metal at the
blade's edge (note how blade straightens out
as it is forged).

Forging the blade

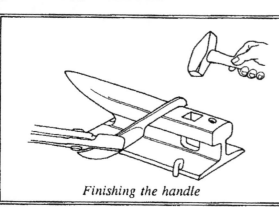

KEY POINTS

Hold correctly in tongs (to avoid blade being
damaged and knife jumping out of tongs).
Blade can now be filed sharp.

Finishing the handle

KEY POINTS — Hardening and tempering

Use two-heat method.
Do not heat above red heat.
Quench out in oil or mud.
Clean back to bright metal.

Temper by heating against a large piece of pre-heated metal allowing heat to move from blunt edge of knife towards blade.
Re-quench at correct temper colour (purple).

FURTHER TRAINING

Next session we are going to make a sickle.

MAKING A SICKLE

DURATION 4 HOURS

OBJECTIVE

By the end of the session each participant will have made a sickle from a length of 10-12mm mild steel bar.

MATERIALS REQUIRED

FOR DEMONSTRATING

Two suitable lengths of 10-12mm round section mild steel bar.
Sample of finished product.

PER PAIR OF TRAINEES

Two suitable lengths of 10-12mm round section mild steel bar.

TOOLS AND EQUIPMENT NEEDED

Hand hammer
Flat-bit tongs
Sledgehammer
Hot set
File or cold chisel
Anvil
Hearth
Bellows
Quenching trough

FINISHED PRODUCT

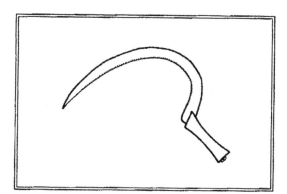

PREPARATION

Have two pieces of mild steel already set correctly in the fire and heating up before the trainees are gathered for the first demonstration.

INTRODUCTION

WHAT

Make a sickle from a piece of mild steel.

WHY

This is one of the products which you can make and sell. Having made this sickle it will be easy for you to make different styles of sickle to suit customers in your area.

HOW

I shall demonstrate each stage, asking some questions, then each of you will make a sickle like this one (hand round sample of finished product).

DEMONSTRATION 1

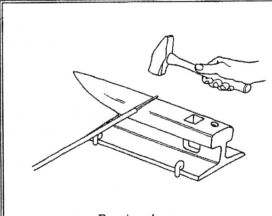

KEY POINTS

Work at correct forging temperature.
Draw down square and bend tang.

Forging the tang

DEMONSTRATION 2

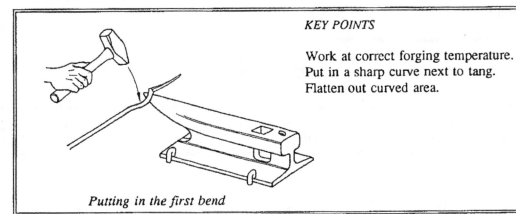

KEY POINTS

Work at correct forging temperature.
Put in a sharp curve next to tang.
Flatten out curved area.

Putting in the first bend

KEY POINTS

Quench hot set every few blows.
Finish cut over edge of anvil.

Cutting to length

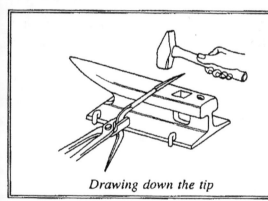

KEY POINTS

Work at correct forging temperature.
Draw down to square taper.
Put in a tight curve.

Drawing down the tip

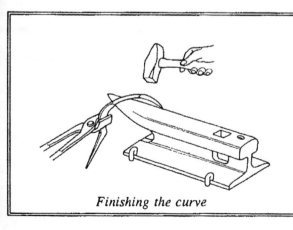

KEY POINTS

Put a tight curve in centre of blade.
Flatten out whole length of blade.

Finishing the curve

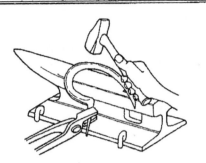

Forging the blade

KEY POINTS

Work along blade evenly, hammering at an angle.
Take care not to burn thin metal.
Sharpen with a file.

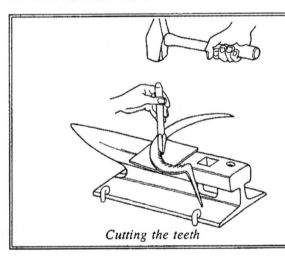

Cutting the teeth

KEY POINTS

Cut teeth either with a file or a cold chisel.
Heat tip of tang, fit handle, bend over tip and quench.

FURTHER TRAINING

Next session we are going to make a traditional pattern hoe.

MAKING A TRADITIONAL ONE-PIECE HOE

DURATION 3 HOURS

OBJECTIVE

By the end of the session each participant will have made a traditional pattern hoe from a used tractor plough disc.

MATERIALS REQUIRED

FOR DEMONSTRATING	PER PAIR OF TRAINEES
Tractor plough disc. Sample of finished product.	Tractor plough disc.

TOOLS AND EQUIPMENT NEEDED

FINISHED PRODUCT

Hand hammer
Sledgehammer
Flat-bit tongs
Hot set
Anvil
Hearth
Bellows

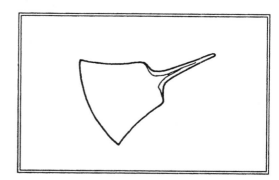

PREPARATION

Have a plough disc already set correctly in the fire and heating up before the trainees are gathered for the first demonstration.

WHAT	Make a traditional pattern hoe from a used tractor plough disc.
WHY	This is one of the products which you can make and sell. Having made this hoe it will be easy for you to make different styles of hoe to suit customers in your area.
HOW	I shall demonstrate each stage, asking some questions, then each of you will make a hoe like this one (hand round sample of finished product).

DEMONSTRATION 1

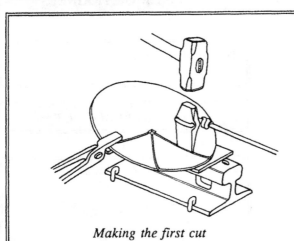

Making the first cut

KEY POINTS

Work at correct forging temperature.
Quench set every few blows.
Cut two-thirds of the way through, allow to cool and break off.

DEMONSTRATION 2

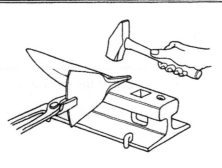

Drawing down the tang

KEY POINTS

Work at correct forging temperature.
Draw down to a square tapered point.

Forging the edge

KEY POINTS

Work at correct forging temperature.
Work evenly along edge.

www.ingramcontent.com/pod-product-compliance
Lightning Source LLC
Jackson TN
JSHW062200130125
77033JS00017B/592